© 2013 Parkinsons Recovery
All Rights Reserved
978-0-9819767-4-7

Language of Recovery

Contents

World of Words 3

Capture Your Own Words 11

Words That Qualify Action 16

Words of Empowerment 21

Conditional Intentions 32

Words of Connection 39

Words that Cancel Intentions 47

Words that Specify Intentions 55

The "Not" Word 61

Words that Negate Intentions to Recover 66

Words of War 69

Words that Affix a Label 75

Words of Retirement 81

Words that Distract 86

Transform Language. Manifest Recovery 92

About Parkinsons Recovery 94

©Robert Rodgers

Language of Recovery

World of Words

"Words are, of course, the most powerful drug used by mankind."

Rudyard Kipling

Looking through my own lens of perception, the world of words and thoughts has shifted over the years. When I was younger, I had the impression people selected their words carefully with the intent to speak the truth.

Of course, people have always lied and distorted the truth over the centuries. That seems to be the human condition. More recently, distortions of the truth seem to have become a way of life in politics, business and the press. Words on their face imply one truth. The underlying truth of words is often the complete opposite of their Webster dictionary meaning.

What is the big deal anyway? Words are just words. You can always take words back. No one is the worse for it. Right? In wars, people get killed with guns. That outcome is terminal. In a war of words, both combatants walk away without a scratch on their body. Or do they?

We have all experienced occasions when the hurtful words of others stuck into our gut like super glue. We can all recall the exact time and place when someone said

Language of Recovery

something to us that was mean and hurtful. Words can certainly harm us and hang around to become bad memories we cannot forget or shake.

Jean Paul Sartre had it right when he said -

"Words are loaded pistols."

We have all been victims and perpetrators of using words as weapons. **Language of Recovery** however sets aside words that do obvious harm to us as well as words we speak from our own mouths that hurt others. [We are all guilty at one time or another!] It is obvious to everyone what language is hurtful.

When others hurt us we know it instantly. When we hurt another person with our own words, it is a challenge to overlook its impact. We know instantly by looking at the shock and pain on the person's face and by their body language. They move away from us as though we had just crawled out from a sewer.

There is however a much more deadly use of words which unfolds for all of us in the way we talk with friends, family and even to ourselves. When we use certain words and language over and over, day in and day out, we can undermine our best intentions to recover from illness and disease. The focus here centers on a much more devious and sneaky weapon – words we use in everyday speech that sabotage our intent to reverse illness. **Language of Recovery** identifies the specific words and phrases that

Language of Recovery

undermine these good intentions and suggests alternatives to promote it.

The pistol loaded with words that undermines the best of intentions is permanently pointed in the same direction. When we vocalize the words that harm - which incidentally can be hundreds of times in one single day – we shoot ourselves in the foot. The pistol is never emptied unless and until we recognize why and how the words we are using in everyday chit chat are killing off our life force and sidetracking the best of our intentions.

Because using the same words over and over becomes so habitual, we are constantly reloading the pistol and aiming it at our feet. We must stop reloading the gun if we ever expect to stop shooting ourselves in the feet. Wounded feet cannot carry us to the destination of our heart's longing. Wounded feet keep us from going anywhere.

There is no doubt about it. Thoughts are at the core of recovery. How do we transform our thoughts so that they facilitate wellness rather than obstruct it? This is a lifetime challenge for each of us. Instead of working with abstract thought forms, **Language of Recovery** tags the words that embrace thoughts that promote recovery and those that impede it.

The words that undermine recovery are subtle in their character and presentation. They are hard to abandon because their ultimate impact occurs so gradually, just like the disease process itself. I will underscore and illustrate

Language of Recovery

words that on their face seem harmless and benign, but which have enduring and devastating effects on success with recovering from disease and illness.

If you are a reader like me, you are likely thinking at this point something to the effect of:

> *"Ok. Ok. I get it. This book is about words I should avoid and words I should use in my everyday speech. Got it. What are they? Get to the gist of it Robert."*

At this early stage of reading the book, you will likely be tempted to fast forward to get a taste of some of the words that are presently undermining recovery. If you have to skip ahead now, please be my guest. But, before you leap ahead permit me to suggest an alternative approach.

Take a snapshot of words you typically use in everyday discussions with friends, family and colleagues. Record responses to incomplete sentences I will suggest next. Pay no particular attention whatsoever to how you express yourself. In other words, use language that is familiar to you before discovering more about whether your favorite words promote your recovery or undermine it.

You can accomplish this task with little effort in several ways. If you prefer to write rather than talk, then write down the responses to the incomplete sentences I have prepared for you to answer. If you prefer talking as a way of communicating, then do an audio recording of your responses. Any answer you give will be a perfect response whether it is short or long and whether it is given in writing or recorded. It is all for your use only.

Some questions may elicit a response that lasts no more than 15 seconds. Others may inspire a longer response of 2 minutes or longer. The length is entirely up to you of course. The point is to capture the specific words that you use by way of custom and habit.

You now know that some of the words you write or speak will be supporting and nurturing your recovery. You also know that some will be undermining it. Feel free to second guess which words help and which hurt. I will warn you in advance that the difference between words that promote health and wellness and those that propagate illness and disease is anything but obvious. Just go for stream of conscious talking (or writing).

How to Make Audio Recordings of Responses to Incomplete Sentences

Perhaps you have a hand held tape recorder you can use for this task. Great. Dust it off and go for it right now. Please do not make a big deal out of this task. Just do it!

Perhaps you do not have access to a tape recorder. Do you have a computer? Perhaps there is a program on your computer you have used in the past. Use that recording program now.

Language of Recovery

Activate whatever method you may have used in the past to record your voice (or write). Dust it off so to speak. Test out the technology. You may need to buy some blank tapes for a tape recorder or update the program on your computer, et cetera, et cetera.

Now, I fully realize you may have never recorded your voice or much less even heard it! If not, your task is to borrow a hand held recorder from a friend or download free recording software onto your computer. If you decide to download the free recording software – here is a link to one system that is reliable and that will insure your computer is free of viruses:

http://audacity.sourceforge.net/download/

Hopefully there is a microphone connected with your current computer. If not, you will have to buy a microphone from a computer store. They are very inexpensive.

Once you download audacity – again, it is free – practice recording yourself talking and then saving the audio file onto your computer. It is important to save the file for listening at a later point in time. Give the recording you make a name like Robert.mp3 (the mp3 is the file extension for the file of an audio recording). If you use a hand held recorder, you will be able to simply listen to what you have recorded which is saved on tape by rewinding the recorder.

Smart phones also allow you to record your voice. They too will work beautifully for the task at hand.

©Robert Rodgers

Language of Recovery

Use whatever recording device is convenient. The old hand held recording devices will work beautifully for the challenges I am about to suggest. If the challenge becomes too much of a burden you will spend your time doing other activities. Make it easy on yourself. Just do it now before you read on.

To summarize -

1. *Choose a recording method.*
2. *Get it up and running.*
3. *Use whatever method is easiest and most convenient for you.*
4. *Don't make a big deal of this task. Just do it.*

I will provide further instructions in the next chapter. Right now, simply get ready and get set to have something that you can use to record your voice.

Maybe you are still thinking – will this activity really be helpful to me? Are words really that important? Please take note of the sign below which was posted in a public restaurant in Washington state.

Language of Recovery

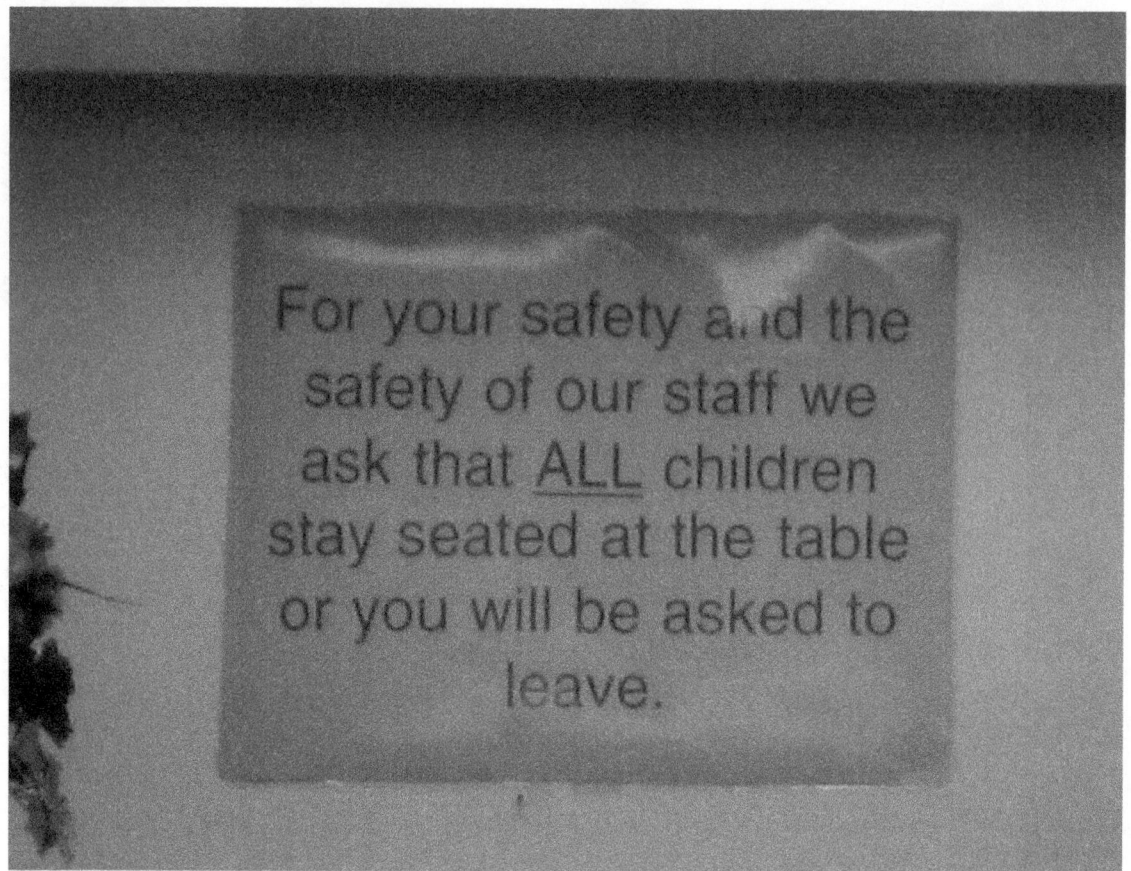

Language of Recovery

Capture Your Own Words

"No matter what people tell you, words and ideas can change the world."

Robin Williams

How do we really know all the many ways we shoot ourselves in the foot? Everyone has a loaded gun filled with words that hurt. Not everyone realizes what the bullets look like. The clues are unveiled through the words we use in everyday language.

The invitation I extend to you (so you can identify what your bullets really look like) is to complete a series of sentences that I will initiate for you. If you have decided to write responses that complete the sentences I am about to suggest, get out your pen and paper now or fire up your word processing program on your computer.

Are you planning to record answers? That would probably be quicker. Have you had an opportunity to set up your recording instrument - whether this is a hand-held recorder or a recording capability that you have access to through a computer or phone? I hope so!

Record (or write) answers to sentences that I will suggest. Do not think about the answers before you respond. Think of this as an ink blot test a shrink is giving you. There are no right or wrong responses. Simply start writing or talking. Allow me to say again - whatever answers you

Language of Recovery

generate are perfect in every respect. If your responses are brief they will be perfect. If they are long they will be perfect. Just allow the talking (or writing) to flow without judgment or criticism.

You are the only person who will listen to the recording (or read your words) unless you decide to allow someone else access . Invite a stream of consciousness to flow. Use words that are familiar and that constitute your every day manner of expression. You will be in a position to derive rich insights when I delve into the explanations of the specific words that help and those that harm.

You want to say the words that habitually come out your mouth. Editing will detract from discovering what words in particular are impeding your recovery. **Language of Recovery** identifies the specific words that will facilitate recovery and those that dampen it.

- *Let there be no censorship in what you write or say.*
- *Let there be no attempt to censor what you think is right necessarily.*
- *Simply talk (or write).*
- *Place thinking on the shelf for now.*
- *Say (or write) what pours out from your mind, heart, body and soul.*

Below is the list of the beginning of sentences which are all incomplete. Your task is to complete each sentence and expand if you are so inclined. You might wish to complete

Language of Recovery

one sentence and then continue on - saying many more before shifting to the next incomplete sentence which awaits completion. Keep in mind that the idea here is to capture the words that you routinely use in everyday talk. The more material you create now, the more insights you will derive later.

At a minimum, complete each sentence. Say (or write) more if you are so moved. Give yourself permission to continue talking or writing for one minute, two minutes, five minutes or even longer if you would like. Alternatively, complete just the one sentence and move on to the next incomplete sentence.

The more uncensored talking that you record, the greater insight you will derive. Feel free to allow the sentence phrases to be a prompt and invitation for you to discuss whatever comes through as you begin talking. No editing is allowed! No drafts are required. Just talk or write!

Finish the sentences below. Capture your words through writing down your thoughts or recording them. Allow yourself to say (or write) whatever comes to your mind without screening it, without stopping and without saying to yourself,

"Oh, I shouldn't be using that word."

If it is coming to consciousness go ahead and voice it or write it! After all – what is the worse thing that can happen

Language of Recovery

once you identify the words that are undermining your recovery? Perhaps indigestion?

"I have never developed indigestion from eating my words."

Winston Churchill

Now finish the sentences below and expand on the thought that is started -

"My recovery is…

"My symptoms are…

"Other people tell me…

"If the truth be told, the prospects of recovery are…

"What I really want in life is…

"Next year I will be…

"Five years from now I will be …

"My family says …

"My doctor says …

"I say …

©Robert Rodgers

Language of Recovery

"Other people with my illness say …

"Other people with my illness think …

"I believe …

"I think …

"I know …

"I want …

"I should …

"My family wants …

"My doctor wants …

"I wish …

"My friends say …

"My true passion is to …

"I love to …

"I fear …

"I predict …

"I see …

©Robert Rodgers

Language of Recovery

Words That Qualify Action

"Words are more treacherous and powerful than we think."

Jean-Paul Sartre

Language shapes our reality. It is a useful mechanism for focusing and directing our energy. It defines boundaries and is accepted literally by the subconscious mind. Of course our egos and our subconscious mind are what actually influence our choices.

What words do you use by way of habit and custom that sabotage your earnest intent to recover? We all shoot ourselves in the feet with the words we choose in one way or another. What words stymie you from taking positive action? Use of any and all words that qualify intentions is one sure fire show stopper.

Words that qualify intentions to recover hold us back from taking clear, positive action. No time is set aside during the day to initiate the actions that are necessary for recovery to unfold. Instead, we slip intentions that have been qualified by our own words into a cue to be addressed at a later time. The problem is that the cue continues to lengthen with each new unattended intention. The cue never gets shorter. Overwhelm becomes a way of life.

Language of Recovery

We nonetheless keep putting more and more intentions in the intention cue, holding out the flimsy expectation that we will eventually have time to take positive action on at least one of them. It is as if we have decided to visit the movie theater. We see a long line and decide to enter the queue at the end. We stand in line, patiently waiting for the line to get shorter. It never does. The line actually gets longer as more and more people break in. We never get to see the movie.

My invitation for you now is to make use of the recording that you just made (or the written document you just prepared). Be on the lookout for any and all words that qualify your intention to recover. You may need to stop the recording you have made to allow sufficient time to write down what it is that you actually heard yourself say.

What are the words exactly that qualify intentions?

- ***You are trying …***
- ***You are hoping…***
- ***You are intending…***
- ***You are attempting to achieve a goal.***

By trying, hoping, intending and attempting … you place all positive action on hold. In so doing, you also place a roadblock on the road to recovery.

- *You are saying 'almost'*
- *You are saying 'possibly'*

Language of Recovery

- *You are saying, 'I think, but probably not quite.'*

There is plenty of wiggle room in the hidden meaning of certain words that stop you from taking actions that are in your best and highest good. Did you "try" to record answers to the incomplete sentences? Did you "intend" to finish the incomplete sentences? If so, I am guessing that you never got around to recording anything. Am I right here? My case rests.

Here is a list of specific phrases that qualify intentions:

> *I am trying …*
> *I am hoping …*
> *I am intending…*
> *I am attempting …*
> *I think perhaps that…*
> *I kind of believe that…*
> *I possibly may be able to…*
> *I almost have…*
> *I would like to be able to…*

Listen now to what you have recorded or read what you have written. How many sentences can you find that contain qualifying words like –

- ***Trying or***
- ***Hoping or***
- ***Intending or***
- ***Attempting or***

©Robert Rodgers

Language of Recovery

- *Thinking or*
- *Perhaps I can or*
- *Kind of or*
- *Possibly or*
- *Almost or*
- *Would like to…?*

Did you come up with a few places where you used these words and terms? Or, did you spot 5, 10 or dozens? It is not uncommon for an individual to spot dozens and dozens of the use of these words with just a five minute recording.

Can you even feel into the limitations of words that qualify intentions? In the sections and places where you have used any of the phrases or words that qualify intentions, write down the context in which the qualifying words were used. This is compelling evidence that you are making a choice that is ineffectual. You are hesitant to take action. The context will connect with a delayed response of -

- *Waiting*
- *Hanging back*
- *Evaluating*
- *Analyzing*
- *Criticizing*
- *Judging*

To summarize - what is the big problem with all of the above behaviors? Aren't we supposed to always be

Language of Recovery

cautious and circumspect about making decisions that affect our health? Of course! But -

- **Waiting honors the symptoms.**
- **Hesitating begs the symptoms to hang around.**
- **"Hanging back" enforces the status quo.**

Waiting, hesitating and "hanging back" work collectively to obstruct the actions that need to be taken for recovery to unfold. The people who insist on analyzing options to death are usually the people who give up on the prospect of recovery.

Language of Recovery

Words of Empowerment

"You can change your world by changing your words... Remember, death and life are in the power of the tongue."

Joel Osteen

Have you had an opportunity to spot all the qualifiers you used in your recording (or writing) that obstruct progress toward a full and complete recovery? Words that obstruct recovery embody intentions that attempt to achieve a goal (like recovery) rather than having accomplished it.

There is a transformation in your use of language that will make a huge difference to the speed and ease of recovery from illness. It helps enormously to replace words that qualify intentions and reinforce thoughts that are not in your best and highest good with words that do support recovery.

I invite you to edit any statements you have made that include qualifying words. Edit all such phrases and statements to reflect a clear, positive intention rather than one which shakes, rattles and rolls - but does little else. Re-engineer the customary way that you talk. Train your highly adaptable neural networks to create new pathways. The body really is more inventive than any human could ever imagine.

Language of Recovery

For example, if you spotted yourself using the word "trying" -

>   *I am trying to…*

You are, in reality - fiddling around with the idea as a child ponders the wisdom of running through a mud puddle. Re-write or re-speak the sentence. Record a different way of saying the same thing:

Instead, now say -

>   *"I am good at…"*

State this intention as something that has already happened. See yourself already in the place that you previously were "attempting" to achieve. Speak from the perspective of having achieved the goal already rather than attempting to achieve it.

Retrospection (looking back on something that has already been accomplished) embodies a position of power and clarity. There is suddenly no longer any question about whether it will happen or not.

People who frame recovery from the perspective of disease and illness unconsciously project the continuation of the same disease state. If they were football or soccer players, they would sign up to be a member of the defense team

Language of Recovery

rather than offense. The defense is an important part of the game, but it rarely scores goals.

 1. *Transform hesitant and qualified posturing to a perspective of power.*
 2. *Assert the reality you want to manifest.*
 3. *Make recovery a reality rather than a fantasy.*

What is the alternative?

- **Be overwhelmed with the many options that will accumulate.**
- **Be inundated with too much information.**
- **Be frustrated with envisioning recovery as a remote possibility.**

Find yourself sitting on the sidelines waiting to play defense in the game of life. The more frequently you use words that qualify, the less often you will even be called in to play a game defense. Instead, you sit on the dugout bench waiting to be called in to play the game. Since you are the coach here and never summon yourself to action, your body withers away as it sits and waits.

Let me get more specific. Take for example the word "hoping."

"I am hoping that I will be able to...

Language of Recovery

Re-write this sentence so that it has already happened.

>*"I am ...*
>*"I now ...*

The word "intend" also dampens enthusiasm for recovery in a flash. Perhaps you wrote or heard yourself say:

>*"I intend to...*

If you intend anything, there likely exists a probability that is typically as high as betting you will not win the lottery. Who makes such a bet anyway? When you "intend" to do anything you sabotage the wish. Substitute other words for "intend" in your speech and talk. You don't want to intend anything.

How about instead saying -

>*I know ...*
>*I see ...*
>*I hear ...*
>*I love ...*

Make your own edits. These are only suggestions of course.

>*"I'm attempting to change my habits of ...*

No, you do not want to be attempting to change any habits. If that is the best you can do, you can be assured your habits will be hanging around for a very long time.

©Robert Rodgers

Language of Recovery

How about -

> *"I change with ease and with grace.*

> *"I am flexible moment to moment, day in and day out.*

Another word that stops action in its tracks is "perhaps".

> *"Perhaps I can ...*

Oops. Again, here is another one of those nasty qualifiers. If this is one of your favorites - it is certainly at the top my list of favorite qualifiers - acknowledge now that the simple use of the word "perhaps" is a roadblock to your successful recovery. "Perhaps" is the key word in the children's book "*The Little Engine That Could*." For those of you who are familiar with this classic, you are well of the struggle the little train encountered to succeed. Why not make it easy on yourself?

Instead of saying -

> *"Perhaps I can ...*

How about instead saying or writing -

> *"I can do anything I set my mind to."*

©Robert Rodgers

Language of Recovery

Instead of saying -

"I can kind of think I will be ...

How about a powerful alternative -

"I am successful."

Another way of saying the sentence above is -

"I am successful, period."

This statement above has a juicy punch to it, eh? No qualifiers are embedded. You are not saying –

"I am successful except for...

You are not saying -

"I am successful, yes. But what I really wanted to accomplish is ...

Can you feel the difference here? The two statements above qualify action, diffuse energy and scramble your mind. They dilute the message so badly that the statement is forgotten in the haze of mixed signals that obfuscate action.

If you spotted yourself saying or writing -

"Possibly I can ..

©Robert Rodgers

Language of Recovery

Re-write or re-speak this statement to -

> "I did it."

Or – if you found yourself saying -

> "I almost can .."

Re-write or speak to :

> "I come from a place of having easily accomplishing all my goals of recovery."

Perhaps you heard yourself say (or your wrote) -

> "I would like to be able to…

No. You really "would" not "like" to be able to do anything. Isn't the more positive statement to acknowledge that you want it now? Then say so.

How about an alternative -

> "I come from a place of having already fulfilled the choices that embody my true passion."

If you hear yourself saying or discover yourself writing in your journal -

- *I am almost …*
- *I am possibly …*

©Robert Rodgers

Language of Recovery

- *I am thinking that perhaps ...*

The intention is diluted and degraded. The real possibility your intention will be realized is instantly reduced to being improbable.

Why is it so improbable you will not realize your intention? Is it because you are not able? Of course not. Is it because it was never realistic to begin with? Of course not. Then why you ask?

I say it is because you shot yourself in the foot by qualifying words that sabotaged your good intentions at the outset. You never even got to first base, much less out of the dugout to play the game. You never even get the chance to try out for the defense team, much less the offense.

All words that qualify good intentions are accompanied by unstated innuendos that there is plenty of wiggle room to lose and little opportunity to succeed. The mind, body and spirit cannot compute or make any sense out of words that qualify.

> *"Sorry Bud. This doesn't compute. I am going on vacation. Call me when you can be clearer about what you really want."*

Two diametrically opposed and entirely incompatible wishes are embodied in words that qualify intentions of recovery.

©Robert Rodgers

Language of Recovery

- *"I am going to do it now" versus "I am not going to do it now".*

- *"Yes, it is good" versus "Yes, it is not good".*

- *"I am going to start now" versus "I am going to stop now."*

- *"I am going to open up and receive healing" versus "I am going to shut down and criticize my healers."*

You are going to say, yeah, I think that may be a good idea but I need more information. Just such a response has happened to thousands of people who have heard guests on my radio show offer a suggestion for recovery that is inexpensive, free, safe and/or has no side effects. People think to themselves –

> *"Hum. That option sounds really interesting. I am thinking that perhaps I should investigating that possibility further.."*

The smart suggestion is delicately placed in a dust pan, only to be emptied in the trash and forgotten the next day. These words have stymied the investigation. Time and energy is withered away. You own words killed your enthusiasm for the idea.

Take the time now to transform the words you use to express intentions for recovery. It will be well worth taking

©Robert Rodgers

Language of Recovery

30 or 45 minutes to re-write any and all phrases or sentences that qualified your genuine intention to recover.

I am quite certain that a large proportion of readers are now thinking this challenge is not worth the effort. Permit me to offer a suggestion that will instantly motivate you to take action now. [Please note my statement "Take Action Now" – with no qualifiers.]

Read the sentence out loud you recorded (or wrote) that qualified your intention. Note how your enthusiasm flops. Then, rewrite the sentence and re-read the edited sentence that contains clear, succinct words. Celebrate how your enthusiasm sizzles from a place of power and enactment. This is the energy that supports recovery from disease and illness.

- *The transformed statement has high power.*
- *It has high potency.*
- *It has high effectiveness.*
- *It has a powerful energetic charge that propels action.*

©Robert Rodgers

Language of Recovery

To summarize – any statement that is embedded with a word or term that represents a qualification creates mental confusion. It frustrates the opportunity to make sound choices from a perspective of clarity. Words that qualify –

- *Convey low power*
- *Diffuse your energy*
- *Scatter attention*
- *Guarantee a lack of focus*

Stamp out those nasty qualifiers today. Speak from the perspective of clarity and conciseness. Celebrate the magic of recovery unfold as your -

Power soars
Energy abounds
Attention centers
Focus is enabled

©Robert Rodgers

Language of Recovery

Conditional Intentions

> *"We are masters of the unsaid words, but slaves of those we let slip out."*
>
> *Winston Churchill*

Conditional intentions confuse the body about the nature and character of your true intention. Several examples of conditional intentions are listed below. Perhaps some will sound familiar to you.

> *"I'm going to recover so that I can visit my grandchildren in California."*

Or,

> *"I'm going to figure a way to dampen my tremoring so that I will not be embarrassed at my Bridge Club on Wednesday."*

Or,

> *"I'm going to visit my doctor next week to see what she can do for me so my wife does not have to keep helping me put my shoes on as if I were a child."*

Or,

> *"Eating foods without sugar is smart because it helps my body heal."*

Or,

> *"I can make myself exercise so that my immune system will be strengthened."*

Language of Recovery

Do any of these statements sound familiar? These are all correct grammatically. These are all certainly good intentions. Certainly these are all ways that you and I have expressed our thoughts at one time or another. There is certainly nothing technically wrong with any of these declarations.

The confound that creates problems is that one intention has been stacked on top of a second intention you also wish to manifest. Both outcomes must happen for the intention to be realized.

Think of your eighty-billion cells as being a bit slow to understand a conditional intention. A cell can understand a clear, singular statement, but when it becomes a conditional statement, confusion abounds.

Alternatively, think of your body as a three year old child. Three year olds understand concise and clear statements. Make it complicated and you lose their attention quickly.

What exactly is a "conditional intention"? What are the signs and signals that a good intention has been diluted by a second condition that is joined at the hip of the first?

Words that signal conditional intentions are words like -

- *"So that ...*
- *"To be ...*
- *"In order ...*

©Robert Rodgers

Language of Recovery

- *"Because ...*
- *"So ...*

For me, "so…" is a huge conditional word I often use. The word "because" is also top on my list. Most people have their own favorites.

Let me now dissect one of my hypothetical examples of conditional intentions.

> *"I'm going to visit my doctor next week to see what she can do for me so my wife does not have to keep helping me put my shoes on as if I were a child."*

Three conditions are embedded in the same statement.

- First of all, the wife has to be in a situation where she doesn't have to dress me.
- Second, I have to visit my doctor.
- Third, I have to wait to see what the doctor can do for me.

Any one of these expectations taken alone is ambitious enough. Join all three together at the hip so to speak and you have declared a herculean task to accomplish. The statement above does not constitute a clear, positive intention that the body is instructed to acknowledge and engage.

Language of Recovery

The body gets confused. It literally loses energy trying to figure and sort out which of these three conditions needs to happen first - if any. Place yourself in the position of your body. What is the body's response?

> *"We are talking about three very complicated circumstances here. I think I will just wait and see what comes down."*

And guess what the body does? It shuts down. Might I add this is a perfectly rational reaction for each and every one of the 80 billion cells – with no exceptions!

The best approach for actualizing good intentions of recovery is to avoid making statements that pile one intention or expectation on top of another. When you see your own good intention has been contaminated by one conditional intention after another, the pile of unfulfilled conditional intentions eventually grows taller and taller. The top eventually touches into the highest clouds in the sky.

Burn that high pile in your imagination. Start fresh. State a crispy clear statement of your most important intention.

How do you clean up the confusion and complications that have been created by a series of conditional intentions that are stacked on top of one another?

©Robert Rodgers

Language of Recovery

Instead of saying -

"So that my wife does not have to dress me…"

(Note the negative in the statement above – "does not have to")

Simply assert –

"I am dressing myself today."

That is the gist of the intention. Just put it out there as such. No visits to the doctor are necessary or required to actualize it. You do need to have an appointment with your doctor.

State the doctor's appointment as an independent activity and just state it as such. Make it simple. Allow that intention to also stand alone.

"Tomorrow, at three o'clock I have an appointment with my doctor."

This is clear. Your "three year old" body gets it. So can you. Add additional clarifications to this intention (if you wish) that all stand alone.

"I'm interested to see what the doctor has to say."

We now have asserted three statements that are independent. They do not co-mingle or mesh with one

©Robert Rodgers

Language of Recovery

another. There is no interdependence. Any one of these statements does not depend on the others.

In what you have recorded or what you have written, scan across or listen for conditional intentions that diffuse your true intentions. It is really not that complicated to spot them. You just have to look for the words that impose a condition on an intention:

> **"So that …**
> **"To be …**
> **"In order to …**
> **"Because…**
> **"So …**

I invite you to tag any and all conditional statements that you made when recording yourself. Disengage the intentions. Make independent statements for each. This is a process which takes time and patience.

I have been working for several months on cleaning up confounds I use in my own everyday language. It has been embarrassing to acknowledge how often I am guilty of confusing my body. It is also rewarding to clean up my language. My body appreciates my hard work, as will yours.

It makes a huge difference to be on the alert for any statements you make that impose conditions on intentions. Clean them up when you hear them spew out of your month, then honor yourself for a good day's work.

©Robert Rodgers

Language of Recovery

To summarize - when the words you use construct a set of intentions that depend on one another, the cells in your body become terribly confused and depressed. The body's processing system (which is far more powerful than any computer) cannot and will not process the complexity. It becomes impossible for the cells in your body hop on board with a clear intention to recover from any illness.

Disentangle and disengage all intentions with have been combined into the same statement. Acknowledge each one separately. Use words that capture each intention that are concise, clear and clean. Outlaw from your everyday language all statements that stack one condition on top of another.

Language of Recovery

Words of Connection

In the course of my life, I have often had to eat my words, and I must confess that I have always found it a wholesome diet.

Winston Churchill

Why do people find that the recovery process from illness is so incredibly challenging? Why does it take so long? Why do so many people have so little success with reversing their illness?

One answer to the puzzle is that the process of recovery is approached on purely a mental level. There is no doubt that significant information can be obtained by using sound analysis and thoughtful research. It makes good common sense to engage research skills to disentangle the many options that are available for recovery from any illness. It is smart to focus on the ones that offer the most promise.

Research and analysis fall into the general domain of mental processes. I love the mental aspect of who we are as human beings. It is the place where I have hung out for most of my life. To maintain health however, accessing only our mental facilities alone does not cut it. Genuine, authentic recovery requires the marriage of our minds with our emotions.

Language of Recovery

Intentions that are purely and exclusively mental in nature do not take us far. Thinking through an intention will nudge us forward in the right direction, but we do not launch a successful program of recovery until we acknowledge and connect our feelings to the intention. When we feel the intention in our body we tap into an entirely different experience than when we merely think it out. When both are in place, intentions materialize swiftly and effortlessly.

Recovery demands that we institute a deep connection with our body and with others. When we disconnect from either, we disconnect from our feelings. When we are numb to our feelings, recovery remains stuck in the mud.

If you are a man and reading this – you may be thinking – Oh my. Where is Rodgers going with this? I have to feel my sorrow and sadness? I have to access my fear, shame and guilt?

No – this is not the point here! You know the feeling of intense excitement that riddles up and down your spine when something exciting is about to happen, right? You know the feeling when something spectacular is about to happen, right? These are the type of feelings that launch recovery into a space of endless possibilities.

Much of what people say is vague and very impersonal. I often hear people talk who use the pronouns "we" or "you" or "our." When I hear these pronouns, I am always speculating on who in the world they are really talking about.

©Robert Rodgers

Language of Recovery

Their spouse?
Their child?
Me?
The population of France?

I often become enmeshed with my own speculations and lose track of what the person is actually communicating. My own thoughts become confused. I suspect their cells are spinning chaotically too (as are mine) with the clutter of mixed messages they are sending.

The magic of recovery requires both the mental aspects and the feeling aspects to be in place. If mental clarity is disconnected from feelings, frustration is likely to mutate into anger. Recovery is halted. If the feeling is present but the mental clarity is lacking, we too will become frustrated and angry. When either exists without the other, the course of recovery becomes haphazard as it winds down a road that spins around in circles.

Everyone disconnects and spaces out at one time or another. Many people (men more so than women) find it challenging to access their feelings. The degree disconnection between the mental level and the emotional level is profound for most people.

There is, however, a simple (and I might add clever) way to unlock the frozen feelings that obstruct recovery. We can do this by recognizing words of disconnection and replacing them with words of connection. I will now

©Robert Rodgers

Language of Recovery

proceed to offer examples that illustrate the sneaky words we use to disconnect thoughts from feelings.

What is your favorite flavor of disconnection?

Consider the statement –

> *"I want to have that life I had always dreamed of having."*

Notice the words "that life." How about making another statement -

> *"I claim my life back today."*

Notice the difference in how the second statement feels when you just read it?

Consider the statement –

> *"I want to be that friend to the body. That is important."*

This is certainly one way to convey your meaning. There is nothing wrong technically with what has just been stated, but I am not addressing technical accuracy here! This statement is impersonal, distant and disconnected.

How about making another statement -

> *"I am my body's best friend."*

©Robert Rodgers

Language of Recovery

This statement has some oomph to it. It has a fierce electrical charge. The body hears this and says –

> *Finally! I have been looking for a friend for a long time.*

Take this sentence of disconnection –

> *"I'm ready to share these gifts that I have, you know."*

Let's make a simple edit to the above statement. Instead of being disconnected and impersonal, why not say –

> *"I always share my gifts with the world, moment-to-moment."*

There is a huge difference in the impact of the second sentence. I feel it. How about you?

Take the following statement of disconnection –

> *"I get all this help from the family. Man. isn't that great?"*

Consider other statements that convey the intention of this thought in a more personal manner -

> *"My husband is always there for me."*

> *"I can always count on my wife."*

©Robert Rodgers

Language of Recovery

Feel into each of the above statements. Contrast the statements above with the depth of the disconnection found in the initial statement. Yes, it is well intended. Yes, the nature of the disconnection is subtle. And yes, is has a chilling effect on accessing your own feelings. The point here is that accessing your own feelings opens the gate to recovery.

Anyone, anywhere could make the next statement -

"I know the family loves me."

Hum … "The family" lives in Slovakia or Moscow or …? And what street was that? I seemed to miss that little detail.

Consider another way of conveying your good intention:

"My family loves me."

Feel that? I do. And because I know you – I also know who the family is and where they live. I know that they do not live in Slovakia or Moscow.

Finally, consider this statement -

"I get the support I need you know."

Language of Recovery

How about another way of saying this -

> *"My friends offer me all the support I have ever needed."*

So you see, there is an energetic charge to each of the statements that we speak and think. Some statements contain vague words that have a weak and ineffectual charge. Other statements contain specific words that carry a powerful charge which connects us to our feelings.

Be mindful of how you disconnect from not only others, but from yourself. There is a lot of disconnection going around these days. It is easy to witness.

When that bond of connection is solidified with each and every statement that you make, your energy will soar. Sometimes you only need to add a pronoun to your statement! The speed of your ability to recover and feel much better with each passing day will accelerate logarithmically.

In summary - become mindful of how you talk moment to moment. You do not have necessarily to examine recordings that you made of yourself talking. You do not necessarily have to dissect a diary that you might have written this week or last year. Simply begin to pay attention to the words that you use to express your thoughts as you have expressed them by way of habit and ritual.
You will surely be able to detect many, many instances of disconnection. I certainly have.

©Robert Rodgers

Language of Recovery

Transform your statements of disconnection and you will transform your program of recovery. You will be surprised at how easy it is to personalize your intentions and pleased with the outcomes that result. This book will be celebrated as the best investment you have ever made!

©Robert Rodgers

Language of Recovery

Words that Cancel Intentions

> *"Don't you know this, that words are doctors to a diseased temperament?"*
>
> *Aeschylus*

At an unconscious level and without realizing it, we often undo ourselves by talking in a way that confuses our body. The intention cancels itself out. Part of us says do it. It is important. The other part of us says, do not do it. I am scared. I am afraid.

Cancellation conveys a statement that implies "Yes" and "No" at the same time. When the body "hears" words of cancellation – and I assure you the billions of cells are listening each and every moment – the reaction is predictable -

> *"Hmm, I don't have a clue what that means. Seems like they are saying yes and no at the same time."*

How does the body react?

> *"Forget it bud. This is not worth paying attention to. I am going to rollover and take a nap until you can get your act together."*

Statements that cancel out themselves can sound rather mysterious. We talk this way so often that we do not even

Language of Recovery

realize or acknowledge the confusing messages we are sending. None of us want to use words that negate and cancel our earnest intention to recover, but we all do it anyway!

Have you ever heard yourself or someone else say -

> *"I'm excited about my prospects for recovery, but my neighbor says no one ever recovers from my type of cancer."*

Can you recognize the cancellation that is embedded in this statement? It comes from the nasty word "but" which contains the letters b-u-t.

Consider making an alternative statement:

> *"I'm excited about my prospects for recovery."*

Period. Your neighbor can think whatever they want. They have their own life to lead. Honor their choice to believe whatever they choose. You have no obligation to accept their beliefs as yours!

Another example –

> *"Won't it be great if I recovered?"*

What cancellation is embedded here? We call this a double-cancellation. The word "won't" conveys the double meaning of both yes and no. Yes, it would indeed be great

©Robert Rodgers

Language of Recovery

if I recovered. Maybe I will. Maybe I will not. The word "if" implies the same negation. Maybe I will recover. Maybe I will not. It is all "iffy" at this point.

Consider an alternative, more powerful statement:

"I am recovering."

Period. Notice the huge energetic difference between the two simple sentences: "Won't it be great if I recovered" versus "I am recovering." The later statement has punch. It has power. It has pizzazz.

Take still another example,

"Hey, doesn't my walking look better today?"

What's the cancellation here? Seems like an optimistic declaration. Right? Maybe – but look closer. You are making two contradictory statements:

"My walking looks better."
"My walking does not look better."

Which one is it? You are asserting both are true when using the word "doesn't."

Why not simply say -

"My walking looks better today. Hooray!"

©Robert Rodgers

Language of Recovery

Another example,

> *"Shouldn't I be asking my doctor about the side-effects of my medications?"*

What's the cancellation? You are saying both -

> *"I should be asking."*
> *"I should not be asking."*

Which one is it going to be? What's a better way of making that statement?

> *"I will talk with my doctor about my medications this week."*

Period. There is no cancellation in this revised statement. It is clear, simple and succinct. Simple is good. Complicated muddles up the intention.

Another example (that I hear often) -

> *"I was wondering, Robert, can't you help me?"*

What are you really saying to me? You are saying: "Robert, I know you cannot help me." And you are also saying "Robert, I know you can help me." You are saying both. In other words, the intention to help is canceled out by the implication that "Robert, you cannot help."

Language of Recovery

How do I respond to such a question? Of course that depends on the situation, but you give me an easy out.

"No my friend. You are crazy to think I can help you. I am not a medical doctor. I am a Ph.D. researcher."

And that is certainly not the response you had bargained for, right?

Why not simply ask -

"Robert, please help me."

There is no cancellation in this simple request. Your request is crystal clear. And by the way, I am certainly much more likely to offer help! My most likely response to the question above is – "How can I help you?"

Let me float another example. You approach your neighbor (who believes there is no chance you can recover from your illness) and ask –

"Wouldn't you like to go with me to the Parkinsons Recovery Summit where they present information about options that are reversing Parkinson's disease?

What are you really saying here? Your neighbor is listening. Your body is listening. What is the translation?

"Go with me."
"Do not go with me."

©Robert Rodgers

Language of Recovery

Do you hear that contradiction in this simple, yet seemingly harmless statement? You are saying yes and no at the very same time in the very same sentence.

What is a much better way of asking? Perhaps you want to ask your neighbor, your wife, your husband, your father, your daughter or your son? Why not look at them and say –

"Please, come with me to the Summit."

This statement is clear. No cancellation is evident in these seven words.

Another example -

"Didn't you love Robert's radio show last week?"

Again, note the contradiction that is implied here.

"You loved it."
"You did not love it."

Your body is listening. Your audience of cells is listening. They are all thinking - "Okay, did I love it or did I not love it?"

How about another way of saying this -

"I loved Robert's show last week."

©Robert Rodgers

Language of Recovery

Period. No cancellation here. Or,

> *"Robert's show last week stunk."*

This too is direct, concise and simple. I do not want to hear it, but I get it.

Be attentive then to the following words that are all clues as to how you will cancel out your true intention. These words are -

- **But**
- **Won't**
- **Doesn't**
- **Shouldn't**
- **Can't**
- **Couldn't**
- **Didn't**

Most of the words listed above are "not" words. Do not miss the most grandiose word of all that we so often use which is **"but."** Listen to yourself talk. You will discover rich insights into the true character of your intentions. Perhaps there is a part of you that does not really want to recover? Your words are clues to the ways you sabotage your own good intentions.

To summarize - when you hear yourself saying words of cancelation stop. Be mindful. When saying yes and no at the same time you are creating enormous stress signals

©Robert Rodgers

Language of Recovery

within every cell of your body. Manifestation of intentions requires that you make clear, unqualified statements throughout the day. When no cancellation is present stress dissolves like a snowball in the sunshine.

Language of Recovery

Words that Specify Intentions

"No man means all he says, and yet very few say all they mean, for words are slippery and thought is viscous."

Henry B. Adams

The quote above refers to men, not women. If you are a woman, did you have a angry reaction when you read it? Am I convincing you about the power of words to evoke feeling? I hope so!

One of the primary ways that language creates stagnation and blocks the process of recovery is through generalities rather than specifics. If we never tie anything down so that it is absolutely clear -

When?
Where?
How?
What?

We can never be held accountable to ourselves or to others. We proudly state a vague, obscure intention without any clear indication of what specific outcome is expected. We refer to a lofty journey, but have no destination. We wander around in circles never getting anywhere. We use

Language of Recovery

vague words that do more to confuse than focus the actions needed to recover.

How often are you guilty of asserting general intentions that have no specificity? We are all guilty of course. Why do we insist sometime on being so vague?

One unconscious reason that underlies such a tendency is it avoids the sad prospect of being disappointed! If the intention is general, you will be unable to track it or even know whether it has happened or not. If you are determined to avoid disappointment at all costs, it is thus rational to frame intentions generally rather than specifically.

In so doing however, you invite the heavy weight of disappointment to ride on your shoulders. The mere anticipation of disappointment will inevitably slow the recovery progress down to a standstill as you travel down the road to recovery. When you avoid clarifying the what, when, why and where of your good intentions, you are unlikely to actualize any of them – ever.

Specifics are the substance that energizes all intentions. Consider some examples of intentions that are so vague and general they lack the potency needed to fire up the actions necessary for recovery to materialize.

"I am going to change my bad habits."

©Robert Rodgers

Language of Recovery

Hello? Which bad habits are you going to change? What ways are you going to change those bad habits? How exactly are those bad habits going to change?

OK. You are going to change your bad habits? Great. Me too. Of course you do not know what my bad habits are, so that means I do not have to change anything. I can say I will change – which makes me look progressive. In the end, I get to retain my same old destructive ways of living. That works for me, eh? I hate change.

Well, maybe I am convinced being vague works for me. It certainly does nothing for my recovery.

Another example –

> *"This week I'm going to do it differently."*

Hmm, do what differently, I might ask? When exactly are you going to do it differently? Is this to be a Monday or a Thursday activity? And, how will it be "different?" If I were one of the billions of cells listening to your words I would now be quite confused about what it is that you truly intend to see happen.

Another example –

> *"It is now clear I need to alter my strategy for recovery."*

Language of Recovery

Hmm, this is a pretty big blob of a statement. What strategy are you talking about? How must this strategy be changed? When are you going to change it, and by the way, who's going to be involved? Are you going to solo this project? Are you going to involve

- *Your cat,*
- *Your horse,*
- *Your dog,*
- *Your wife,*
- *Your husband,*
- *Your daughter,*
- *Your son,*
- *Your grand-daughter?*

Who's going to be involved, if anyone?

Another example –

"I am working on my recovery everyday."

Okay, congratulations. This is wonderful news. How so? What exactly are you doing?

- *When are you doing it?*
- *How often are you doing it?*
- *Does this really always happen every day?*
- *Do you plan on doing it two days every week?*
- *Is all of what you are talking about here work or is perhaps some of what you are talking about fun?*

©Robert Rodgers

Language of Recovery

- *Do you never really even skip a day?*
- *Are you really talking about each and every day?*
- *Are you referring to an eight-hour day or ten-hour day or a twenty-hour day or does your intention refer to eight minutes on Tuesday morning?*

Let's start getting specific with intentions. When you get specific you empower your intention.

Another example –

"I'm dealing with the symptoms."

Great. But I've got some questions for you so we can make that a little more specific. How so?

- *Are you playing cards with them?*
- *Are you resigned that you will always feel this way?*
- *Are you fighting your symptoms?*
- *Are you creating a wall between yourself and your symptoms so that you are no longer listening to the signals your body is sending to you?*
- *Are you numbing your symptoms so that they can totally and completely be ignored?*

Another example -

"WOW. This week I am starting to feel a difference."

©Robert Rodgers

Language of Recovery

Well, this is wonderful news but please tell me more. You are feeling a difference in what exactly? And, by the way, is the difference in a positive direction or are you feeling worse? After all, a difference can be in either a positive or negative direction. Are you feeling better or worse?

When you say "start," do you mean for sixty seconds at 2:34 PM? You are not aware of a bothersome symptom? Or, did you enjoy your dance class tonight so much that you experienced no bothersome symptoms for the entire two hours of that fun activity?

To summarize - generalities stagnate good intentions. Listen more carefully to the words you use when you make statements of recovery. Catch yourself when the statements and words are too general. Then elaborate. Tease out the specifics.

General statements lead you into the desert with no water to drink. Specific statements propel you forward in the direction of recovery as prosperity, health and abundance pour into every cell your body. Specifics are the fuel that organizes the actions necessary to manifest recovery

©Robert Rodgers

Language of Recovery

The "Not" Word

"Words are like weapons. They wound sometimes."

Cher

How often do you use the word "not" in everyday speech? I did a word search on the "nots" that I used in writing this book and found a ton of them. Little did I know I was such a negative guy! Are you too more likely to express what you do not want to see happen than what you do want to see happen?

When we use the word "not" our intent is to express something that we do not wish to see happen. Statements of exclusion are clear and succinct. You now wonder what is the problem my friend? In actuality, we are insuring that the opposite will be enacted. When we vocalize the wish "not" to see something happen, it increases the likelihood it will happen!

When we say what we do "not" want, we give credence and credibility to what it is that we do "not" want. When saying that we are -

- *Not being ...*
- *Not doing ...*
- *Not having...*
- *Not wanting ...*

©Robert Rodgers

Language of Recovery

- *Not desiring ...*

The universe of consciousness responds quite logically -

"Okay. Got it. I will pay attention to that now."

You see, the word "not" is not comprehended or "heard " by the universe which only knows abundance. When the word "not" is used, we insure that more focus, energy and attention is now directed to what it is that we do "not" want.

The subconscious mind also does not comprehend, "hear" or acknowledge the word "not". It does hear and acknowledge and notice what it is that you do *not* want to see happen. Let me provide some examples of what I heard myself saying in my thoughts and also in my words recently.

"I do not want to have problems paying the mortgage next month like I did this month."

My intention next month is clear, isn't it? I want it to be easy to pay the mortgage. But my subconscious computes my intention very differently than I had intended. It says –

"Hmm, seems like he enjoyed not being able to pay the mortgage on time. I guess we'll just have to give him that experience of frustration again."

©Robert Rodgers

Language of Recovery

When the subconscious sees the image below (and of course it does see!) the response is simple: "OK. Guess it is time to turn on the panic button."

Be conscious of each and every time you hear yourself thinking or saying the word "not."

1. **Press the pause button.**
2. **Evaluate exactly what you intended.**
3. **Rephrase your statement so that it no longer includes the word "not".**

©Robert Rodgers

Language of Recovery

I will provide several examples to explain the process above.

Suppose you hear yourself thinking or saying -

"I do not want to struggle this week."

Instead, say -

"My movements this week will be effortless."

Or, consider a statement you have most certainly said at one time or another –

"I choose not to do that again."

How about adopting an alternative statement,

"I choose to learn from my mistakes."

You will surely notice when you hear yourself using the word "not." Stop. Pay attention to the meaning you had intended. Acknowledge you are never going to manifest your intentions when you state them in negative terms.

Be easy on yourself. Everyone does it! Then, erase the statement and replace it with a positive statement that no longer includes the word "not".

©Robert Rodgers

Language of Recovery

To summarize - using the word "not" when you express your intentions does exactly the opposite of what it is that you intend to see happen. Notice each and every time you use that nasty word "not." Restate your intention in a positive frame which is the foundation that is essential for recovery to manifest.

Language of Recovery

Words that Negate Intentions to Recover

"We die in proportion to the words we fling around us."

Emile M. Cioran

The word "not" is not the only word that undermines intentions. The following words also negate good intentions:

- **Stop**
- **Quit**
- **Never**

When we use these words, just like when we use the word "not", we are sending that sneaky subconscious mind of ours mixed signals. The confusion which results empowers precisely what we do "not" wish to see happen.

Some examples will help explain how I suggest that you become more mindful of using these three words. Perhaps you have said to yourself recently -

"I am choosing to stop this struggle."

The subconscious mind hears the word struggle. The reaction is to say -

©Robert Rodgers

Language of Recovery

"Hmm, struggle? They must like to struggle. I guess we'll have to make sure that continues."

As an alternative – I suggest that you say –

"I choose to move with ease and grace."

Another example –

"I choose to quit eating sugar. I know it's making me sick."

That certainly is an ambitious and important intention. Your subconscious mind is going to be thinking -

"Hum, the only thing I really understood was the word "sugar". I certainly know what sugar is. I will now make sure that we pay close attention to any and all foods that contain sugar. Guess they want more of it. I can handle that request with ease."

How about expressing your intention in a different way -

"I choose to claim my healthy habits of eating nutritious food."

A final example -

"I am choosing to never do that again. I know that missing a day of exercise is not in my best and highest good."

©Robert Rodgers

Language of Recovery

How about an alternative –

> *"I choose to exercise whenever it serves my best and highest good."*

Do you see the difference? Of course you may now be thinking -

> *"Well, shoot, you are just splitting hairs here. You are saying the same thing."*

It is certainly true that on the mental plane – both statements imply the same thing for the most part. With any statement we send signals to our subconscious mind that sidesteps the mental realm. After all, the function of the subconscious is to derail the mental level. When we use words that are clear, direct and clean, the subconscious gets it.

To summarize - words that negate intentions slow recovery down to a standstill. Jump Start your program of recovery by reframing any and all thoughts and statements that negate your good intentions. You do have to be more mindful of the words you use when you speak but the benefits are enormous. This program of recovery is also free.

Language of Recovery

Words of War

"Good words are worth much, and cost little."

George Herbert

There is no doubt about it. People have different reactions when they are told by their doctor that they have a disease. Some people go into a serious, deep and profound depression. Others react by thinking -

"Okay, I'm sure there has to be a medical doctor out there who has a solution for me."

They then make inquiries to the best medical institutions in the United States, Canada and across the world to locate a medical resource that has a cure. For some people that strategy works beautifully. For others, it does not.

Other people go into resignation. They say to themselves, -

Oh, this must be my fate. I have been wondering what will happen to me. Now I know."

They choose to resign themselves to whatever is going to unfold for them. No positive action is taken whatsoever.

Language of Recovery

Still other people say to themselves,

> *"I know there have to be dozens of options out there, options that I can pursue and consider that can reverse the symptoms that I am currently experiencing."*

This third response is one that I endorse and advocate through my work with Parkinsons Recovery.

There is also a fourth response that is adopted my many.

> *"I'm not having this. I'm going to fight this battle tooth or nail. I am going to be the victor. This is a war that I am going to win."*

What do you say to your family and friends about the symptoms that you currently experience, whatever they might be? Do you hear yourself saying -

> *"This is a battle you know. It is a war. I am here to tell you I will prevail. I will be victorious."*

In your thoughts and in your words, be mindful of the specific terminology that you use when you think about what is happening with your body when you tell others about the symptoms that you have been experiencing. What words do you use?

©Robert Rodgers

Language of Recovery

- **Have you adopted the word "war"?**
- **Have you latched onto the term "battle "?**
- **Is this a World War which pits you against your body?**
- **Have you declared a war just as if you were a soldier on a battlefield?**
- **Are you on one side of a conflict battling a ferocious enemy?**
- **Who is the enemy?**

If you do use these terms, what energy surges inside your body whenever you say them? When I hear them, I become aggressive and hostile. My adrenaline flows like Niagara Falls. My suggestion is to eliminate the use of any such words.

What happens in a real battle I ask you? History shows us that soldiers on both sides clearly die. In some wars millions of soldiers have died on both sides of the battle. Millions of civilians die too - innocent individuals who were in the wrong place at the wrong time: women, children and men who do not wear uniforms.

During World War II the Allied Forces bombed cities in Germany indiscriminately, killing millions and millions of innocent German citizens. Perhaps you would say they deserved to die because no German citizen was innocent? Germans were killing innocent people too. Perhaps the civilians deserved to die because they allowed Hitler to

©Robert Rodgers

Language of Recovery

assume power. These are the rationalizations used to justify the bombings at that time.

If you have declared war on your body, you want the bad cells (the cancer cells or the infected cells or …) to die and the healthy cells to thrive. This is not what happens in the body with most of the treatments used for cancer today. The end result is no different than the bombing of Britain by the Germans or the bombing of the Allied Forces in Germany. There is a fall out of millions of innocent victims.

When war is declared on any disease you murder with your words billions of the living, vibrant cells, organisms and tissues that are the primary means your body uses to rebound from any illness over the long term. With many cancers, for example, radiation is used which does thankfully succeed in killing the cancer cells. Radiation also destroys the healthy cells that are needed to nurture a healthy immune system. Without a healthy immune system, the body does not have a prayer of coming back into balance.

OK. You are not convinced. You insist on declaring war on your disease. Who is your enemy? Is your enemy your neurons or heart or lungs or liver? They are all certainly doing the best job that they can do under the circumstances. Who and what are you fighting?

Think of it this way. If this is a real battle, then the enemy has to be your own body. You are literally pursing a

©Robert Rodgers

Language of Recovery

strategic program of words to murder yourself. Is that what you really want to do? I don't think so.

When you engage war like energies, you activate the cortisol and the adrenalin in your body. That turns the switch on to high gear. To win a war soldiers cannot take a break. They must -

Go .. Go … Go …
Move … Move … Move …
Fight … Fight … Fight …

The minute you get up in the morning and the minute you go to sleep, soldiers must always be on high alert. This is certainly not a successful approach for healing and recuperation. Doesn't the body need rest to recuperate? I certainly always thought so.

For healing to succeed, there needs to be a gentle combination of excitation and relaxation, an effortless flow of energy running through the meridians of the body. Ongoing life cycles of "turning it on" then "turning it off" are required for the cells to rejuvenate and sustain life.

To summarize - be aware of the consequences when you declare war on any illness. Of course, you do want to problem to be solved. Of course, you do want to see these symptoms resolve. The problem is the war like terms that you are using activate an energy that is not in your body's best and highest good.

©Robert Rodgers

Language of Recovery

The healing response is to acknowledge and to honor whatever symptoms you may be experiencing in the moment and begin asking the questions -

"What's here now? What does my body need that I am not giving it?

- *Do I need to exercise?*
- *Do I need to move?*
- *Do I need to eat different foods?*
- *Do I need to stop eating certain foods?*
- *Do I need to have more social contact?*
- *Do I need to acknowledge my true purpose in life?*
- *Do I need to change jobs?*
- *Do I need to use gentle methods to help my body detox?*
- *Do I need to get bodywork so that I can begin to release the trauma that is stored at the cellular level in my body?*
- *Do I need to begin loving myself?*

What is that I need to do for my body in the moment? This is the healing response which does not and cannot unfold on a battlefield.

©Robert Rodgers

Language of Recovery

Words that Affix a Label

"Sticks and stones will break our bones, but words will break our hearts."

Robert Fulghum

If you are reading this book, you or a loved one has most likely been diagnosed with a disease or illness of one type or another. Perhaps the condition is cancer or MS or Parkinsons or … Whatever the disease label might be, how many times have you said a three word sentence that explicitly names the disease?

- **"I have Parkinson's."**
- **"I have cancer."**
- **"I have MS."**
- **"I have …"**

How do you feel when you say these three words? How do you feel when you think it? Do you spot any of these three words in the recording you make at the outset of this journey into the world of words that help and those that hurt?

What is the big deal here you are wondering?

1. You have gone to the doctor.
2. The doctor has ordered a series of tests.
3. The doctor diagnosed you with a disease.

©Robert Rodgers

Language of Recovery

4. You are simply stating a truth about the diagnosis your doctor has determined.

Here's the big deal. Any word that labels a disease state connects to a mass of dark, negative energy. Permit me to focus on one disease in particular – Parkinson's disease. Many people who have been diagnosed with Parkinson's disease hold the belief that the condition is degenerative and progressive. They are convinced that their symptoms will get worse and worse over time.

My research through Parkinsons Recovery shows this belief to be unfounded and blatantly false. Anywhere up to 50% of individuals who have been formally diagnosed with Parkinson's disease have been misdiagnosed. How can this be so? It certainly has nothing to do with the competence of medical doctors. They get extensive training and their qualifications for the most part are irrefutable.

The challenge for doctors responsible for making a diagnosis is that there is no definitive test to determine whether or not a person has Parkinson's. It's really a call based on experience, guess work and intuition. It should come as no surprise then that many guesses (that translate into diagnoses) turn out to be wrong.

Medical doctors do have to make a diagnosis for any condition or else they cannot prescribe medications. If you have decided to try medications or surgeries, doctors must diagnose first before any treatments can be administered.

©Robert Rodgers

Language of Recovery

Some diagnoses embody a large set of symptoms which are unexplainable. A surprisingly large number of diagnoses are simply a label for a condition that is really not well understood. Such diagnoses are better known as "garbage can" diagnoses. Why affix a label to yourself that does not have any clear causal origin and is unreliable at the outset?

Perhaps the diagnosis is right on target. It is also likely that other imbalances are also present in your body which is why a disease state has emerged in your body.

When you declare -

"I have [name of disease] ...

You may well be defining a tiny part of the problem. To be honest about it all, to declare all of your imbalances could include a frightening list –

"I have liver damage"
"I have kidney stones"
"I have high blood pressure"
"I have bones that are calcium deficient"
"I have ...

How much negativity can you tolerate in a lifetime? Who wants to declare a long list of complicated conditions? Who wants to hear it? Why bother going to the trouble to memorize everything that is out of balance in your body? Isn't that what doctors are paid to do?

©Robert Rodgers

Language of Recovery

I suggest the following. Let doctors do the diagnosing. Their job is to diagnose. They can place a label on your symptoms. That is their determination and their responsibility. The label need not be yours to treasure, pin to your chest and announce to the world.

I have some suggestions for you to consider for how to talk about the disease or illness you may be currently experiencing.

Some people say -

> *"My doctor has diagnosed my symptoms as Parkinson's disease."*

That is very different from saying

> *"I have Parkinson's disease."*

There is a deep resignation to the declaration above. It is as if you have fallen deeply into that false belief template that Parkinson's is degenerative and progressive.

I think it is useful to describe what happens to be going on with your body. If, for example, you have neurological challenges and want to tell somebody what really is going on with you, you could say something to the effect of,

> *"I'm experiencing some stiffness in my body and a bit of internal tremoring."*

Language of Recovery

Use words that describe what you are personally experiencing. Use words that describe what is real for you in the moment. Make them descriptive, factual and true for you.

If you say –

"I have Parkinson's disease.

You invite in a low-mo state of energy. Names attached to diseases are nouns, not verbs. Nouns describe conditions that are set in stone. Your body however is changing every minute. If you use a noun (in the name of a disease) to label yourself, you endorse the thought form that Parkinsons is degenerative. You have made a declaration that you do not expect to recover. You plant a doormat at your doorstep that says – "Depression Welcome." Why do you want to put yourself in a position of predicting a degenerating health condition which is not a valid prediction in the first place?

When it comes to how you talk about any disease, I recommend that you simply scratch the statement -

"I have [name of disease] ...

Eliminate this three word sentence from how you explain your situation to family and friends. Everyone has cancer cells. I do not announce to my friends

©Robert Rodgers

Language of Recovery

"I have cancer."

But I do have cancer cells. Most people have a neurological challenge of one type of another - but they do not go around telling their friends they have MS or ALS or Parkinsons Disease.

Find other ways to describe what is going on with your body. Your optimism and attitude about recovery will transform quickly if you catch yourself whenever you label yourself as a disease.

> ***You are not a disease. You are a person. Medicine takes the person out of the equation. That is why so many people are open to options these days.***

To summarize - when you state that you have a disease – "I have …" it is possible that you may not actually have the disease you thought you had. You are certainly not doing yourself any favors when you declare a condition that may not even be true in the first place. Catch yourself whenever you label yourself as a condition rather than a person.

Find other honest and constructive ways to express what it is happening with your body in the present moment. Allow your words to flow from your heart. You will know they are the right words because it will feel good when you say them. You will also know because recovery will begin to unfold gently and effortlessly.

©Robert Rodgers

Language of Recovery

Words of Retirement

"Why should I say I will retire in three or four years? You retire the very moment you utter those words."

Haile Gebrselassie

Words I hear frequently in conversations are "retire" and "retirement."

"After I get my 25 years in, I am going to retire. Only 244 days left."

"George is enjoying his retirement after 30 years with the company."

My mother bought my father a present when he retired from being a patent lawyer – a wide screen TV. His life consisted of sitting in his Big Boy chair in front of the television until he needed full time nursing assistance. His final years were spent in a nursing home that locked the residents in because they would walk out and get lost if the doors were unlocked.

When I decided to quit academic life, my school announced they were giving me a retirement party. I had only worked at the University for 14 years. I explained I was quitting, not retiring but they insisted on calling it a retirement ceremony. Faculty members chipped in to award me the customary retirement present – a clock.

©Robert Rodgers

Language of Recovery

At the ceremony I thanked my colleagues for their thoughtfulness in sponsoring the retirement party and for the clock.

> *"Now I will have something to do. I can watch the clock tick all day long."*

My joke did not take. They thought I was being serious.

The word "retirement" triggers a strong, negative reaction in me. Why? Consider the definitions of "retire."

> *"To withdraw to a secluded place..."*

After my "retirement" I am moving to the desert and will no longer be in contact with anyone? I think not.

> *"To go to bed ..."*

After my "retirement" I am going to bed and stay there for the rest of my life? I think not.

> *"To retreat ..."*

After my "retirement" I am looking forward to retreating from manifesting my passions? I think not.

> *"To give up one's work ..."*

©Robert Rodgers

Language of Recovery

After my "retirement" I will abandon all of the skills I have acquired during my lifetime and play marbles? I think not.

"To cause to retire from a position ..."

After my "retirement" I will sue my employer for forcing me to retire from the job I loved? I think not. I quit!

What positive connotation does the word "retirement" convey?

- *Does the word "retirement" suggest transition to a life that promises excitement, adventure and new opportunities? No.*

- *Does the word "retirement" suggest a change in circumstances that fulfills a person's passion more closely? No.*

- *Does the word "retirement" suggest an opportunity to engage work that fills a person's heart with joy and satisfaction? No.*

I suggest that the word "retirement" be "retired" from your vocabulary as it has mine. Many people look forward to their "retirement" only to discover that they become ill and need constant medical care. They took too seriously one meaning implied by the word "retirement" to go to bed. Their true intention was to begin living. Their words expressed the opposite intention.

©Robert Rodgers

Language of Recovery

Many people tell me they were forced to "retire" because of their illness. Might I suggest other words that convey a higher order intention?

> *"I am excited about changing my circumstances this month. This will give me more time to engage my passion in life which has always been ..."*

What words do I use today to reflect on the occasion of leaving academic life which seduced me for 14 years with job security and a secure salary?

> *"My decision to quit academic life in 2003 and pursue my true passion of researching causes and treatments for Parkinson's disease launched the most exciting adventure of my life."*

When I hear the word "retirement" I instantly become depressed. The word itself sends chills up and down my spine because it conveys the approach of imminent death. I want to embrace my life today. I am not interested in wasting away in bed, watching television while sitting in my Big Boy Chair or sitting out in the hot desert all by myself.

To summarize - I respect the choice of other people who embrace and celebrate their "retirement". I even honor their choice of words. I for one make a different choice. The word "retirement" has been deleted from my vocabulary. I chose an alternative path to embrace and celebrate my new life that allows me to devote my energies

Language of Recovery

to doing what is in my best and highest good rather than spending my time spacing out in faculty meetings.

Language of Recovery

Words that Distract

"Think twice before you speak, because your words and influence will plant the seed of either success or failure."

Napoleon Hill

Most people use certain words frequently that distract from the true meaning they intend to convey. The most common utterance used in the English language is "uh." The "uhs" that are uttered outdistance words of profanity. When you say –

"*uh …*"

Or when you say -

"*So…*"

Or when you say -

"*What I mean is…*"

You add nothing to what you want listeners to hear. More importantly, you distract them from understanding the true meaning you had intended to convey. Listeners take your ideas less seriously when you inject words like –

- **So**
- **You know**

Language of Recovery

- **Uh**
- **Hum**
- **I think**
- **Well**
- **Yeah**
- **But**
- **I would guess**
- **What I mean is**

Catching yourself using distracters can be totally and completely fun and, at the same time, totally and completely frustrating. Begin noticing and acknowledging any and all distracters that you customarily use when you talk.

When I began hosting the Parkinsons Recovery Radio Show years ago it took some months to become familiar with how my voice sounded to me when I edited the recordings. What I heard was not what I was accustomed to hearing when I talked. The true surprise as it turns out was the many, many qualifiers that I used when I talked. Quite frankly, I was shocked.

I began counting the number of "uhh…uhh…uhs" that I used when I talked and was flabbergasted. I challenge you to count the number of distracters you use when you speak. Count the number you used when you recorded yourself talking. If you count them now, I predict you will be just as shocked as I was.

A second invitation which has the promise to be entertaining is to engage a family member or friend in

©Robert Rodgers

Language of Recovery

helping you become aware of distracters that you use in everyday speech. Ask them for a day, an evening or an afternoon to catch each and every distracter that you utter. Of course you could ask them to be very specific and to alert you as to when you use a very specific distracter like "uh." Or, you could simply ask them to raise a finger every time you use any particular phrase that they believe distracts from what you have said.

Alternatively, ask them to keep a tally. Instead of alerting you each time you utter a distracter, ask a family member or friend to make a secret count. For example, as you begin dinnertime they could keep a tally of the distracters you utter.

Another possibility you may want to consider is to make more recordings of yourself talking - even in casual conversations - and then listen to the recording afterward. If you use Skype – you can make recordings of yourself while talking on the phone. It is likely you will discover (just as I did) that you do use many distracters when you talk.

Distracting terms are unnecessary words and utterances that add nothing to your true intent. Include in this challenge any and all profanity words that add nothing to the point that you are intending to make (unless of course they help to emphasize your meaning!). Some people use a word of profanity in every other sentence.

How many times have you listened to a person talking who used so many qualifiers that you began thinking -

Language of Recovery

"Can't you just get the point? I really can't wait another 5 minutes for you to say the words I need to hear to understand what it is that you are actually saying."

When we use distracters in speech our thoughts are not taken seriously. People in the public arena practice talking so that they don't use utterances that distract. They well know that their audience will stop listening after 15 seconds if their statements are packed full with words like "uh," and "well" and "what I mean is…" If a politician can learn how to eliminate distracters from their speeches and debates, anyone can learn to do it, eh?

There is another, much more important reason to become mindful of distracters that you inject when you talk. You are using these same qualifiers when you think about the possibility of recovery. I, as you know, interview hundreds of people every year about their thoughts with regard to recovery.

I often ask the question -

"So, how is it going? How is it looking for you?"

The answers that I hear are clues about the real intent the person has set for recovery. I will now paraphrase some of what I have heard over the last six years:

"Well, you know, the doctor says that it's really not going to be possible, so of course, you know, I'm going to him for a long time and I know, um, that well, he's you know, pretty prestigious guy, he's, you know,

Language of Recovery

well, he's at a really, I mean, uh, a really good university and I, well you know, I--I really got the best man, and--and, that I could possibly get. And so, I don't know, I just hope that uh, well, I don't know, I just hope that down the road that uh, I guess if I could just not get--get worse, I mean, you know. You know that would be really--that's I'm like, well, like, I mean that's what really, that's what I guess in the end I'd like--I'd like to see."

Maybe you think this is an exaggeration, but I assure you it is not. When people respond, there are usually a multitude of distractions that are embedded in how they express themselves which obscures their true intentions.

Contrast the convoluted response to the question above with the following response-

"I'm recovering every day. Hooray!"

End of story. That is it folks. I got it. You got it. We all heard it. That is all there is to the answer. That is the gist of the intention. No distracters or qualifying terms and needed or necessary. There is absolutely no doubt in my heart, mind and soul and yours about what is meant by the five word sentence above.

Let's switch back to a response with a different flavor that also insures recovery will be stuck in the mud -

"Well, yeah, you know, um, well, the way it turns out is I think all things considered, probably, you know, well, what people – what a lot of people really say is

©Robert Rodgers

Language of Recovery

that they, they think I'm probably better, that I'm recovering and, you know, I don't know, I don't know exactly, well, I'd say, yeah, I'd probably, yeah, I'm recov--I do have, well, like everyone I have, you know, I have, I have bad--some bad days here and there, but I guess I'm--I'm probably, yeah, I'm probably recovering. I--I guess, yep, I suppose that's right."

The contrast between the clear and clean statement of recovery and the two confounded and convoluted statements is stark. Statements loaded with reservations and hesitations invite listeners to stop listening. The cells in the body stop listening too.

Set the intention to take whatever action is needed to reverse symptoms and become symptom-free.

1. **I am ready to live my life.**
2. **I am ready to actualize my passion.**
3. **I am ready to activate my life force.**

To summarize - it helps enormously to eliminate any and all distracters when you talk to anyone about anything. And, of course, as you begin to notice and eliminate words that distract from your meaning - as you talk out loud to others – these expressions of doubt and hesitation are the very same distracters that you use when you think about reversing symptoms. Clean up your language and you clean up the thought forms. Crisp language expressed by clear words is the fuel that empowers recovery. Take delight in the end result.

©Robert Rodgers

Language of Recovery

Transform Language. Manifest Recovery

"How true Daddy's words were when he said: all children must look after their own upbringing. Parents can only give good advice or put them on the right paths, but the final forming of a person's character lies in their own hands."

Anne Frank

The words we use to express our thoughts are crafted more from habit and custom than thoughtful reflection. Words can be weapons or treasures. We choose words to include in each and every statement that we make to our loved ones, our friends and to ourselves as thoughts rattle around our minds.

I personally have not been attentive to how I phrase my thoughts when I think as well as when I talk with others. I access the same old, well paved neural pathways in my body to retrieve the same words again and again. My choice of words is made out of habit and ritual.

Some words and phrases that I use by way of habit undermine my true intent. I know that now, after working on this book! A benefit of writing ***Language of Recovery*** has been to identify all the words I use that sidetrack my own good intentions.

Language of Recovery

My invitation for you is identical to the invitation I extended to myself. Be more mindful of the phrases and words that you use by way of habit and custom when you talk and write. Catch yourself saying words that obstruct your intention to recover as they pop out of your mouth. Rest assured they will. Edit your statements in real time. Celebrate the transformation that empowers your recovery.

When we talk, think and write it is true that our words flow from a singular entity. Words are organized in our brains to frame statements. We spit them out in words from our mouth (some of us more frequently than others). The illusion is that no one else is listening when we think our words silently to our self. There may be other persons listening when we vocalize statements to others, or others may read our written thoughts.

Think of the impact our words have from a new perspective. Each of us is the residence for seventy or eighty billion cells, far more than the population of the earth. Each cell has a consciousness of its own. Whenever we express a thought silently to ourselves or express an idea out loud to another person, we have an audience of some seventy or eighty billion listeners.

Since cells have a consciousness, they can easily become confused about our intentions. They can have discussions among themselves about what in the world are "we" (the cells) are supposed to be doing now. They too can become confused, just as we become puzzled about what we are doing or where we are going with our lives. Cells obviously

©Robert Rodgers

do not have mouths to speak their thoughts, but there is a consciousness present. How else would they be able to sustain life?

Each sentence we craft, each statement that we make, each word we speak either obfuscates our true intent or clarifies it. Our audience of eighty billion (give or take a few cells) is always listening. They listen when we use words that clearly convey the idea. They are turned off when we use words that are obscure and misleading.

Your thoughts are precious to others and to yourself. Treat them as such. Your audience of cells thanks you for using words that convey intentions clearly and succinctly. Your body will thank you profusely.

About Parkinsons Recovery

Robert Rodgers, Ph.D. is the founder of Parkinsons Recovery which provides ongoing support, resources and information to persons who currently experience the symptoms of Parkinson's disease. The programs, radio shows and services provided by Parkinsons Recovery are also accessed through the main website –

http://www.parkinsonsrecovery.com

www.ingramcontent.com/pod-product-compliance
Lightning Source LLC
Chambersburg PA
CBHW051351200326
41521CB00014B/2536